EXTREME PLACES

The
Longest
Tunnel

Other books in the Extreme Places series include:

The Longest Tunnel

Kelly Borchelt

KIDHAVEN
PRESS™

THOMSON
＊
™
GALE

San Diego • Detroit • New York • San Francisco • Cleveland
New Haven, Conn. • Waterville, Maine • London • Munich

© 2004 by KidHaven Press. KidHaven Press is an imprint of The Gale Group, Inc., a division of Thomson Learning, Inc.

KidHaven™ and Thomson Learning™ are trademarks used herein under license.

For more information, contact
KidHaven Press
27500 Drake Rd.
Farmington Hills, MI 48331-3535
Or you can visit our Internet site at http://www.gale.com

LIBRARY OF CONGRESS CATALOGING-IN-PUBLICATION DATA

Borchelt, Kelly.
    The Longest Tunnel / by Kelly Borchelt.
      p. cm. — (Extreme Places)
Summary: Discusses the construction and history of Japan's Seikan Tunnel, the longest and deepest tunnel in the world.
Includes bibliographical references and index.
    ISBN 0-7377-1882-X (hardback : alk. paper)
    1. Seikan Tunnel (Japan)—History—Juvenile literature. 2. Tunnels—Japan—Design and construction—History—Juvenile literature. [1. Seikan Tunnel (Japan)—History. 2. Tunnels—Japan.] I. Title. II. Series.
    TA807.B67 2004
    624.1'93'0952—dc22
                                                        2003015617

Printed in the United States of America

Contents

CHAPTER ONE

Closing the Gap

The Seikan Tunnel of Japan is the longest tunnel in the world. It stretches 174,240 feet, or almost thirty-four miles. It would take a car traveling sixty miles per hour more than thirty minutes to drive from one end of the tunnel to the other!

The Seikan is a railway tunnel. It connects Japan's main island of Honshu with its northern island of Hokkaido. Almost one-half of the tunnel, or about fourteen miles, passes under the water of the Tsugaru Strait. The rest cuts through the earth under the jagged coastal hills on both islands.

In addition to being the world's longest tunnel, the Seikan is also the world's deepest underwater tunnel. The Tsugaru Strait is over 450 feet deep, and the top of the Seikan Tunnel is another 330 feet below the seabed.

That places the tunnel nearly 800 feet below the surface of the water. At that depth, you could stack the Statue of Liberty on top of the tunnel twice and it would still not be visible above the water.

Plenty of Problems

When the Seikan Tunnel was completed in 1988, it offered a quick and safe route across the Tsugaru Strait. But it took more than forty years for this to happen.

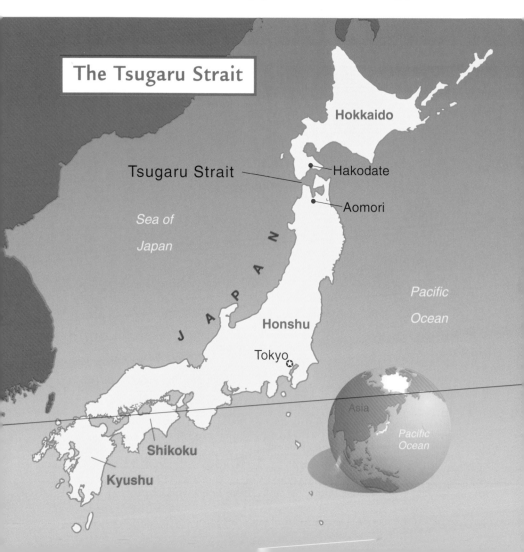

The Tsugaru Strait

In the 1940s, the population of Japan was growing rapidly. The large island of Hokkaido promised thousands of square miles of open land. The island also had valuable resources, such as fertile soil, cold-water fishing, and coal. However, it was very difficult to get there because there was no quick, safe, affordable way to cross the strait.

Ferries had long served as the main form of transportation between Honshu and Hokkaido. The ferry ride across the strait took several hours and was fraught with danger. Thick fog, sudden storms, and savage tropical hurricanes called **typhoons** often battered ferries and their cargo of people, cars, and goods. One especially vicious storm in 1954 sunk a ferry, killing more than thirteen hundred people. The Japanese people were enraged by the horrible accident and demanded that a better answer be found for travel between the islands.

Dangerous Traffic Jams

In addition to severe weather, heavy ship traffic also made crossing the Tsugaru Strait risky. Many cargo ships from neighboring countries passed through the strait on their way to and from the Pacific Ocean. Dangerous traffic jams resulted from the large number of ships heading through the strait. Ships, both large and small, crisscrossed the waters of the strait like cars in the parking lot of a busy mall. Collisions were common.

People file past coffins of the victims of a ferry that sunk in the Tsugaru Strait in 1954. Authorities needed help identifying the more than thirteen hundred dead.

Some travelers chose to fly between the islands, but this option was not very popular. Flying cost too much money for the average person. At that time, airlines were not as common as they are today, and airplane tickets were too expensive for such a short trip.

Only One Solution

A tunnel was not the first, or only, idea for connecting Honshu and Hokkaido. One plan was to build a bridge between the islands. However, even at its narrowest point, the strait was so wide that the world's longest bridge would have been miles too short to span the distance. Bridge construction would also have been nearly impossible in the turbulent conditions of the Tsugaru Strait. Even if a bridge could have been built across the strait, the violent typhoons and heavy fog would have made crossing it at least as dangerous as ferry travel.

Beneath the Strait

Of all the ideas considered, an underwater railroad tunnel seemed to make the most sense. Traffic jams and air pollution caused by cars had already led to plans for railroads on Japan's mainland. A railroad tunnel beneath the strait could connect to this network of trains. However, the idea had some risks.

The biggest challenge was the ground beneath the strait. A tunnel would have to run through a swirl of rock and soft soil along an earthquake zone with nine **fault lines**. At any time, any one of these large cracks in the earth could shift, dangerously churning the ground all around the tunnel. However, the **engineers** believed that the earth around the tunnel would also act as a cushion, protecting it from earthquakes once it was complete.

Studying the Earth

Before planning could begin, engineers and **geologists** had to study the earth under the Tsugaru Strait. They needed to know what conditions workers could expect

The idea of building a railroad beneath the Tsugaru Strait made the most sense, because it could be connected to Japan's existing railway system.

hundreds of feet below the surface of the strait. Since no other tunnel had ever been dug so deep, they also needed to prepare as best they could for the increased risk of flooding and cave-ins. In 1946, a geological survey was started to provide this information. Digging, however, was still a long way off.

The Competition

Although many underwater tunnels were already in use throughout the world, most of them ran under shallow rivers. None were as long, or as deep, as the Seikan. Up to this point, the world's longest underwater tunnel was Japan's Dai-Shimizu. Once completed, the mam-

The World's Longest Tunnels

Name	Year Completed	Location	Length
Seikan	1985	Japan	33.5 miles
Chunnel	1994	UK/France	31 miles
Moscow Subway	1990	Russia	23.6 miles
Chesapeake Bay	1964	U.S.	17.4 miles
Iwate-Ichinohe	2002	Japan	16 miles
Dai-Shimizu	1976	Japan	14 miles

British and French workmen celebrate the construction of the Chunnel.
At thirty-one miles long, the Chunnel is three miles shorter than the
Seikan Tunnel.

moth Seikan was more than twice the length of the
fourteen-mile-long Dai-Shimizu.

Today, the Seikan Tunnel's closest competitor is another underwater tunnel. The Chunnel runs under the
English Channel, a large, choppy body of water between England and France. The underwater part of the
Chunnel is almost nine miles longer than that of the
Seikan. However, when the Chunnel was finished in
1994, its total length was about three miles shorter than
that of the Seikan Tunnel.

13

At Long Last

The Seikan Tunnel solved the problem of getting people across the Tsugaru Strait faster and safer than ever before. However, the process of building the world's longest, deepest tunnel ran up against several problems. Many obstacles had to be overcome to finish such a unique structure.

Getting Started

The construction of most tunnels is simple. The work begins at one end and moves forward until the workers reach the other end. Construction of the Seikan Tunnel was more complicated than this. It had to be built in stages.

Three Is Better than One

Engineering plans for the Seikan called for three tunnels. There would be one large main tunnel for train travel between the islands. Two smaller tunnels, to be used by workers during and after construction, would not be as long or as tall as the main tunnel. These tunnels, known as pilot and service tunnels, would be about half as tall as the three-story main tunnel and only about a third as long. Before construction could begin on the main tunnel, the two smaller tunnels had to be dug.

The Seikan Tunnel project officially began with the digging of the pilot tunnel, started in Yoshioka, Hokkaido, in 1964. Geologists used the pilot tunnel for their surveys. These surveys helped them to determine the condition of the earth around the main tunnel.

Digging on the service tunnel started in Tappi, Honshu, two years later. The service tunnel provided workers with ventilation and a path for taking materials in and out of the main tunnel.

In most tunnel projects, dirt or **spoil** from digging is carried out through an opening that also serves as an

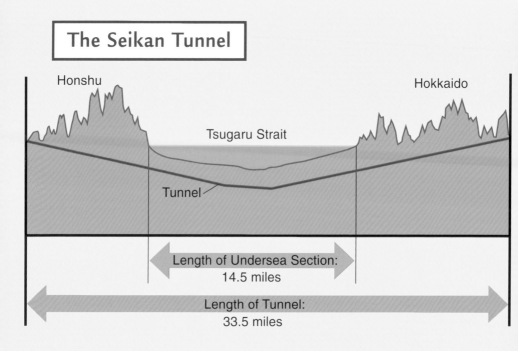

The Seikan Tunnel

Honshu

Hokkaido

Tsugaru Strait

Tunnel

Length of Undersea Section:
14.5 miles

Length of Tunnel:
33.5 miles

entrance for workers. The Seikan's service tunnel saved time by allowing workers to move tools and debris through one opening and enter through another. The two tunnels were connected by several tiny tunnels called galleries. These were dug between one thousand and twenty-five hundred feet apart. In this way, workers going in and out of the tunnel did not disrupt workers who were drilling and blasting in the main tunnel.

Driving the Tunnel

Since tunnel construction always moves forward in one direction, like a car traveling down a street, the digging is called tunnel driving. During the time that the Seikan Tunnel was being planned, advanced technology had given workers a great new tool for driving tunnels. The **tunnel boring machine (TBM)** made tunnel construction faster, easier, and safer than older methods. A TBM is like a giant can with dozens of sharp cutting blades on one end. It turns slowly as it pushes forward through the earth. As it moves, the blades spin like giant saws, cutting and scraping away the surface of the new tunnel. Some of these machines are massive. The one used to drive the Chunnel was as tall as a three-story building!

Unfortunately, geological surveys turned up a mix of soft, wet soil and rock around the Seikan Tunnel site. This soggy mix could not be carved out by a TBM. Another technique, known as **sinking**, was also ruled out.

Many tunnels are constructed using a tunnel boring machine like this one. A tunnel boring machine could not be used for the Seikan Tunnel because of the soft, wet soil and rock at the site.

Project managers relied on a combination of old and new methods to construct the Seikan Tunnel, seen here near completion in 1985.

Sinking involves building large sections of a tunnel above ground, then sinking them in a body of water, where they are welded together. The strait proved too deep for this method.

The strait's depth prevented the use of other common tunnel-driving techniques. A buildup of pressure was the main concern. Too much pressure would create a safety hazard for workers.

Despite these challenges, tunnel construction pushed forward. Project managers had to rely on a mix of old methods and some new ones designed especially for conditions below the Tsugaru Strait. Older, riskier methods of drilling, blasting, and **mucking** (or removing soil from the tunnel) had to be used. Newer tunnel-building techniques such as **shotcreting** were put to use for lining the tunnel's inside walls.

Because drilling had to be done by hand, it took a very long time. To begin, workers drilled carefully placed holes in the earth at the front of the tunnel. These holes were then filled with explosives and blasted apart. Workers then carried the chunks of rock and soil out through the service tunnel.

Once several large blocks of earth had been blown apart and removed from the tunnel, a tunnel shield was put up. The tunnel shield protected workers from a collapse until the tunnel walls could be supported by concrete. As the blasting progressed, the tunnel shield was moved forward.

Once the workers were well into the tunnel, they started work on shotcreting. To build a permanent lining, the workers constructed a type of skeleton made from steel girders. They placed this around the tunnel walls. The girders were then sprayed with a thick layer of cement that dried into a smooth wall of concrete. When complete, the Seikan Tunnel consisted of almost 170 tons of steel. That is enough steel to build five of New York City's Empire State Building!

Slowly but Surely

Knowing the Seikan Tunnel would take a long time to build, engineers searched for ways to save time. One way they did this was to start each of the three tunnels at both ends. The crews worked their way toward a center point under the strait. This meant that there was

A worker checks the tunnel wall during drilling. Drilling by hand is a risky construction method that takes a long time to complete.

work being done on six different tunnel sections at one time. The tunnels stretched out like three long fingers from each island until they finally met in the middle, far below the Tsugaru Strait.

Driving the tunnels in this way required years of work from the engineers before any digging could be done. Thousands of complicated measurements had to be taken to make sure the tunnels would meet in the same place deep underground. Even the smallest mistake might cause the tunnel ends to miss each other, delaying the project for months or years.

Workmen inspect the interior of the Chunnel, which connects Great Britain and France.

Construction workers use machines like this one to dig tunnels. Unpredictable soil often slowed digging in the Seikan.

The unpredictable soil at times slowed the work. In one area, the rock was so difficult to dig through that it took the workers four months to dig only forty feet. It took an average of one month to dig a distance of barely ten feet! In another spot, the soil was so soft and mushy that the workers had to inject the muck with cement. This made the ground hard enough for drilling.

Challenges like these ate up a lot of time. It took almost seven years before the pilot and service tunnels could be used to start work on the main tunnel. Even after all that time, the two smaller tunnels had not been finished, but they reached far enough underground for the workers to begin driving the main tunnel. This meant that the pilot and service tunnels were being used while they were still being dug. Once the main tunnel was started, another seventeen years would pass before the entire project was finished.

Finishing the Job

When digging was finally started on the main tunnel in 1971, the Seikan project had already been in development for over twenty-five years. As the engineers had expected, there were many challenges to building the world's longest and deepest tunnel. The work was slow and dangerous. Some problems were unpredictable. Others were common to tunnel building, especially with underwater tunnels.

Danger in the Tunnel

One of the biggest hazards of building underwater tunnels is pressure. The deeper the tunnel, the greater the pressure of water and soil pushing down on the tunnel's surface. Since no other tunnel in the world was as deep as the Seikan, the water pressure was greater than had ever been encountered on any other tunnel project. All that pressure threatened floods, **blowouts**, and a deadly disease known as the **bends**.

The risk of flooding was the biggest threat to the Seikan project. Nearly eight hundred feet of soil and water put intense pressure on the tunnel walls. Over the years, four major floods and several cave-ins disrupted work on the Seikan. Protecting the tunnel workers from these unpredictable dangers was a constant challenge.

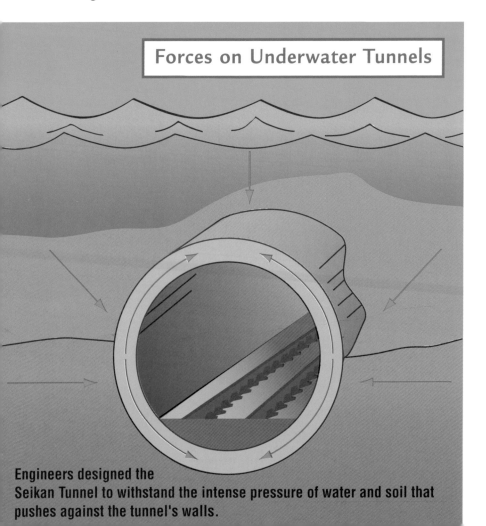

Forces on Underwater Tunnels

Engineers designed the
Seikan Tunnel to withstand the intense pressure of water and soil that
pushes against the tunnel's walls.

Flooding was the biggest threat during construction of the Seikan Tunnel, seen here in the 1970s. Four major floods and several cave-ins disrupted work during the seventeen years of construction.

One of the worst disasters that occurred during the project was a major blowout in 1976. Pressure from the earth and water above the Seikan punched a huge hole in the tunnel wall. Water gushed into the tunnel at a rate of forty tons a minute. It took the workers over two months to control the massive flooding. All digging in that section of the tunnel stopped during that time. Amazingly, no workers died in this accident.

The Bends

Another hazard was the bends, a disease common to workers in deep tunnels. It is very painful and can be deadly. All humans have a gas in their blood called nitrogen. The body usually absorbs the nitrogen easily. But when a worker deep underground moves to the surface too fast the nitrogen forms bubbles. The bubbles can expand and pop like the spray from a shaken soda can. This can cause serious injury. To protect the Seikan's workers from this disease, they worked in the tunnel for only one hour of every eight-hour day. Then, each worker spent several hours in a **decompression** area before leaving the tunnel.

Earthquakes

As if the pressure were not enough of a risk, the Seikan was also dug along several fault lines in the middle of an earthquake zone. The engineers expected the finished tunnel to be safe from earthquakes because the earth would cushion it during any violent shaking. However, the workers digging the tunnel risked being trapped or crushed if an earthquake struck during construction. The unfinished tunnel walls might collapse under the pressure and bury the workers under rubble and a rush of water.

Other dangers involved the explosives used in blasting. Cave-ins sometimes resulted from blasting. Cave-ins

To avoid the bends, workers on the Seikan Tunnel only worked one hour a day.

could lead to drowning or cause a worker to be crushed by huge chunks of soil and rock. Or a worker might be suffocated or poisoned by toxic fumes released into the air of the tunnel.

These and other problems challenged the engineers and workers like few other projects. Yet somehow the death rate from accidents remained low. Between the time the pilot tunnel was started in 1964 and the entire project was finished in 1988, thirty-four workers died. With 14 million people working on the tunnel over the life of the project, this number was remarkably low.

Meeting in the Middle

The final blast was set off in the Seikan's pilot tunnel on January 27, 1983. The completed pilot tunnel connected the islands of Honshu and Hokkaido for the first time in history. A short time later, the service tunnel was finished. Finally, after more than two decades of construction, the Seikan Tunnel was almost complete.

On March 16, 1985, the two ends of the main tunnel met in the middle and the digging was finally done. However, the workers were not finished just yet. It would be several more months before construction on the railway was completed. It would also take some time to clean and test the rail systems that would operate in the tunnel. Finally, in March 1988, the first commercial rail service began in the world's longest tunnel.

Almost 14 million people, or one-quarter of Japan's workforce, had worked in shifts of three thousand at a time, twenty-four hours a day, to build the Seikan Tunnel. Even with thousands of people working night and day, twenty-four years had passed since the first shovelful of dirt had been dug out of the pilot tunnel.

Workers celebrate the completion of the Seikan Tunnel in March 1985. Construction and testing of the railway inside the tunnel would take another three years.

A construction trolley glides along the completed rail system in the Seikan Tunnel. Commercial rail service began in March 1988.

In the end, the project also cost much more than anyone ever dreamed. When it was being planned in the early 1940s, the Seikan Tunnel engineers expected the project to cost almost $600 million. By the time it was finished, the tunnel had cost almost $7 billion. That was ten times the original budget!

Despite all the problems it encountered over the years, when the first trains began rolling through the tunnel in 1988, the Seikan project was still seen by most as a marvel of modern engineering.

The Tunnel Today

The Seikan Tunnel has been open and operating for more than fifteen years. Today the tunnel is used for both passenger and freight transport between Honshu and Hokkaido. About 2 million passengers travel through the Seikan every year. There are also up to fifty-two freight trains running back and forth in the tunnel each day.

Travel in the Tunnel Today

At one time, tunnel planners hoped to see a high-speed bullet train running through the Seikan Tunnel. For various reasons, this did not come to pass.

The main passenger train that runs in the Seikan Tunnel is called the Kaikyo, which is the Japanese word for strait. Each day the Kaikyo makes seven trips between the cities of Aomori on Honshu and Hakodate on

Hokkaido. Since the train runs on the slower, standard rail, the trip takes almost three hours. That is nearly twice as long as it would take if the trip were made by a bullet train. The same trip takes only a little over an hour by airplane.

There are also two sleeper trains, the Hamanasu and the Cassiopeia, that both make one daily round-trip. These special trains are very popular. They are designed for passengers that want to make their ride in the Seikan a vacation. The sleepers are an inexpensive, easy way for the local Japanese population to get away from home and enjoy a short trip with their families.

Once the Seikan Tunnel was complete, the smaller pilot and service tunnels were put to other uses. To-day these tunnels serve mostly as maintenance and

Seikan Tunnel Facts

Location: Honshu and Hokkaido, Japan
Year Started: 1971
Year Completed: 1988
Cost: $7 billion
Length: 174,240 feet (33.5 miles)
Purpose: Railway
Type: Underwater
Materials: Steel, concrete
Engineers: Japan Railway Construction Corporation

Construction workers board a train in the service tunnel of the Seikan in 1983. Today, the pilot and service tunnels serve as maintenance and drainage areas.

drainage areas. If disaster were ever to strike in the Seikan, the pilot and service tunnels could also be used as escape routes.

The maintenance depots are another part of the Seikan Tunnel that are not being used for their original purpose. These depots were designed for maintaining bullet trains. Unfortunately, the cost of construction ran so high there was not enough money to pay for the expensive tracks needed for the bullet train. Instead, the depots are on display, like exhibits in a museum, for tourists visiting the Seikan Tunnel. Also popular with tourists are two undersea train stations at either end of the tunnel.

A bullet train arrives in Kyoto, Japan. Bullet trains are not used in the Seikan Tunnel because money is unavailable for the special tracks needed for bullet trains.

A Bittersweet Success

The tunnel successfully met a large part of its main goal. Travel across the Tsugaru Strait is now safe from the dangers of frequent storms, and a greater number of people can be transported across the strait at one time. However, the Seikan Tunnel has not lived up to expectations.

The tunnel's main purpose was to cut travel time between Honshu and Hokkaido. Japan's Shinkansen, or bullet train, is the fastest train in the world. If it had been used, travel between the islands would have been very quick. With the slower, traditional train now in use, the trip between the islands is shorter than by ferry but much longer than hoped.

The increase in the popularity of air travel has also hurt the Seikan railway. When the tunnel was being planned, flying across the Tsugaru Strait was so expensive that few people picked that method of travel. However, over the years, airlines were able to lower their ticket prices, and flying from Honshu to Hokkaido became much more affordable and popular.

Almost since it opened, the Seikan railway has had trouble competing for passengers. Travel in the tunnel has dropped by a third, or about 1 million passengers per year, since it was opened in 1988. Despite the popularity of the two sleeper trains, the number of passengers that ride them is not large enough to make up for the decrease in the overall number of tunnel travelers.

Today, the affordability of air travel between Honshu and Hokkaido has hurt business on the Seikan railway.

The Seikan's future is uncertain. Too few passengers could mean too little money to operate the railway and maintain the tunnel. In addition, the Seikan Tunnel is aging. Many of the tunnel's metal pieces, like bolts and steel beams, are beginning to rust after so many years deep underwater. Repairs are made as often as possible,

The Seikan railway has declined in popularity since its opening in 1988, and its future is uncertain.

but most of the maintenance projects are very large and expensive.

The Seikan Tunnel remains the longest tunnel in the world. How long it will continue to be a working tunnel is unknown.

Glossary

bends: A disease caused by high pressure that commonly affects workers in deep underground tunnels.

blowout: A burst of rock, soil, and water through a hole in the wall of a tunnel.

decompression: The process of relieving pressure.

engineer: A person responsible for planning, constructing, and managing a project.

fault line: A large crack in the earth's crust.

geologist: A person who studies the origin, history, and structure of the earth.

mucking: The process of removing moist, sticky soil from a tunnel.

shotcreting: Spraying the walls of a tunnel with cement to form a concrete lining.

sinking: The method of driving a tunnel by sinking sections of the tunnel into water and then welding them together.

spoil: Rock, muck, and spoil cut from the walls of a tunnel.

tunnel boring machine (TBM): A large cylinder that pushes through the earth, chipping away rock and soil with sharp, spinning blades.

typhoon: A tropical storm common in the Pacific Ocean.

For Further Exploration

Books

Andrew Dunn, *Structures: Tunnels.* New York: Thomson Learning, 1993. Read about how tunnels are built and learn some interesting facts about the Seikan Tunnel. Includes craft projects.

Sam Epstein and Beryl Epstein, *Tunnels.* Boston: Little, Brown, 1985. This book describes the many dangers of tunnel driving and includes a short history of the Seikan Tunnel and some of its competitors.

Gail Gibbons, *Tunnels.* New York: Holiday House, 1984. A brief introduction to tunnel types, shapes, parts, and building methods.

Stephen Hoane, *The World of Caves, Mines, and Tunnels.* New York: Peter Bedrick Books, 1999. An illustrated explanation of how caves, mines, and tunnels are used for exploration and travel under the earth's surface.

Chris Oxlade, *Bridges and Tunnels.* New York: Franklin Watts, 1994. An illustrated explanation of how bridges and tunnels are built.

Websites

Discovery Channel Extreme Engineering (www.discovery channel.com). This page includes links to exciting stories about current and future tunnels, as well as other extreme structures.

Encyclopedia Encarta Tunnel Article (http://encarta. msn.com). This article explains the different types of tunnels and tunneling methods. Includes facts about some of the world's greatest tunnels.

PBS Wonders of the World Databank (www.pbs.org). This site includes a list of facts and stories about the Seikan Tunnel. There are also links to other pages with information about tunnels and their construction.

Index

Picture Credits

About the Author

Kelly Borchelt is a freelance writer who enjoys traveling and studying the world's most amazing man-made structures. She earned a bachelor's degree in marketing from Texas A&M University in Corpus Christi, Texas. Borchelt resides on the Texas coast with her husband, Preston.